Snowdonia Metal Mines

Recording the past

Des Marshall

Gwasg Carreg Gwalch

First published in 2022
© text & photos: Des Marshall

All rights reserved. No part of this publication
may be reproduced, stored in a retrieval system,
or transmitted in any form or by any means,
electronic, electrostatic, magnetic tape, mechanical,
photocopying, recording, or otherwise, without prior
permission of the authors of the works herein.

ISBN: 978-1-84524-462-0
Cover design: Eleri Owen

Published by Gwasg Carreg Gwalch,
12 Iard yr Orsaf, Llanrwst, Wales LL26 0EH
tel: 01492 642031
email: books@carreg-gwalch.cymru
website: www.carreg-gwalch.cymru

Contents

Introduction ... 5

1. **ABERLLYN LEAD AND ZINC MINE and COED MAWR POOL LEAD MINE**
Aberllyn SH 794 579 and Coed Mawr Pool SH 779 584 ... 10
2. **ABERGLASLYN and BRYNFELIN COPPER MINES**
Aberglaslyn SH 594 462 and Brynfelin SH 589 472 ... 14
3. **BRAICH YR OEN and HAFOD Y LLAN COPPER MINES**
Braich yr Oen SH 614 516 and Hafod y Llan SH 623 523 ... 16
4. **BRITANNIA or SNOWDON COPPER MINE**
SH 616 547 ... 20
5. **BWLCH Y PLWM LEAD MINE**
SH 626 413 ... 22
6. **CEFN COCH GOLD MINE and the associated BERTHLWYD and NEW CALIFORNIA GOLD MINES**
SH 717 234 and Berthlwyd SH 722 235 ... 24
7. **CLOGAU GOLD MINE**
Old Clogau SH 678 198; St David's lode SH 676 202; Clogau No 2 SH 673 201; Lefel Fawr, Ty'n y Cornel SH 672 201 ... 28
8. **CRIB DDU or LLWYNDU COPPER MINE**
SH 606 482 ... 30
9. **CWM BYCHAN COPPER MINE**
SH 602 472 ... 34
10. **CWM CIPRWTH** SH 525 477 ... 36
11. **CWM DWYFOR COPPER MINE**
SH 541 506 ... 38
12. **CYFFTY LEAD MINE**
SH 772 588 ... 42
13. **DRWS-Y-COED** SH 543 534 and associated mines **SIMDDE DYLLUAN** SH 543 533 and **BENALLT** SH 536 535 ... 44
14. **GARTH GELL GOLD MINE**
SH 688 200 ... 46
15. **GLASDIR COPPER MINE**
SH 739 225 ... 48
16. **GWYNFYNYDD GOLD MINE**
SH 735 282 ... 50
17. **HAFNA LEAD MINE**
SH 780 602 ... 54
18. **HAFOD Y PORTH COPPER MINE**
SH 780 602 ... 56
19. **KLONDYKE LEAD MINE**
SH 765 621 ... 58
20. **LLANBERIS COPPER MINE**
SH 597 587 ... 60
21. **LLANRWST LEAD MINE**
SH 779 593 ... 62

22. LLIWEDD BACH COPPER MINE SH 634 531 — 64	**28. PARC LEAD MINE including VALE OF CONWY LEAD MINE** Parc. Lead. No 2 Adit SH 786 598; No 3 Adit SH 787 601; No 5 Adit SH 788 607; Fuchelas Deep Adit SH 786 600; Air Saft and Stopes SH 787 596. Vale of Conwy. Lead. SH 778 596 — 78
23. LLYN BODGYNYDD SH 764 595 — 68	
24. LLYN DU MANGANESE MINE SH 654 346 — 70	
25. LLYN DYWARCHEN MANGANESE MINE SH 657 341 — 72	**39. SYGUN COPPER MINE** SH 605 487 — 80
26. LLYN EIDDEW MAWR MANGANESE MINE SH 645 340 — 74	**30. YSTRAD/SILURIAN or GARREG FAWR IRON MINE** SH 544 579 — 82
27. PANDORA LEAD MINE SH 767 602 — 76	**Glossary of Terms** — 86 **Selected Bibliography** — 88

Ruins at Hafod y Porth

Introduction

When one thinks of mining and quarrying in Snowdonia the mind instantly thinks of slate. This of course was the major industry in the area and it was said that Welsh slate 'roofed the world'. Much less known about are the metal mines. These, too, greatly contributed to the economy of the area. The main metals found in Snowdonia are lead, copper, manganese, gold and iron. Gold was found and mined during 'gold rushes' in great quantity. Welsh gold was the gold of choice for Royalty. On Moel yr Ogof, a satellite peak of Moel Hebog, there is a trial white asbestos mine. The Romans were probably the first to recognise the potential wealth of metals to be found with lead being discovered and mined by them. The main mining activity began around 1550 but virtually ceased in the 1930's although some mines continued for much longer, even until the late 1990's.

As with slate mining and quarrying the legacy of the mineral mines is slowly disappearing whether this is occurring from natural causes or vandalism. My *Snowdonia Slate, the industrial heritage of a National Park* recorded what can be seen on the surface today before it disappears. Carrying on with that theme I wanted to record what can be seen of the metal mines before they too are lost to brambles, bracken and trees which are taking over and hiding these wonderful places.

There are still many remnants of metal mining in the area and these are no less spectacular than those of slate. Remains at Hafna, Parc, Klondyke, Glasdir and Cefn Coed for example provide a great insight into the struggle in extracting the mineral wealth. Workers often lived in barracks during the 6 day working week often in great hardship. The distance the miners walked to hew these metals from the ground gives an indication of the desire and determination of the miners to find metal deposits.

Metal mining has long been a major industry in Snowdonia, especially so from the mid 19th century to the early 20th. Similarly to the lead lodes in the Conwy Valley, copper ore is found in the Ordovician rock in natural faults. Copper is often found alongside lead or galena as it is properly called. The copper mines in Snowdonia are smaller than the one found around Parys Mountain, Anglesey. This resembles a moonscape and is a wonderful place to visit. Between 1804 and 1931 the main mines produced around 35,000 tons of copper. Drws-y-coed was the most productive at 13,000 tons. This was followed by Simdde Dylluan with 8,000 tons. Both these are to be found in Dyffryn Nantlle (Nantlle Valley) near to Tal-y-sarn. Llanberis mine produced 7,500 tons, Britannia or Snowdon 3,000 and Sygun, close to Beddgelert 1,500.

Iron in limited quantities has been mined with the major one, Ystrad, close to Waunfawr in the bedded oolitic or pisolitic rocks of the Cambrian or Ordovician period. (Ordovician was termed after a war – like tribe of Celtic/Welsh people called the Ordovices before the Roman era). The diagonal slash running up the hillside is a remarkable feature and has numerous levels and some very deep shafts. Iron was deposited in the sea during sedimentary depositions and are not in hydrothermal veins. The maximum output was around 80,000 tons in 1857.

Lead has been mined since the 16th century and there were a plethora of small workings. Many of these were joined up to become quite major concerns with only a few being financially viable, such as Aberllyn and Parc. The most productive period for many of these mines was during the second half of the 19th century when some 25,000 tons of lead and zinc were produced mainly from the area around Llyn Sarnau.

Lead is found in steeply dipping faults along three distinct directions: North to South: East North East to West South West and East South East to West North West. The broadest veins were usually found in the North to South veins and were often 80 feet wide whilst others were much narrower and only 6 feet or so wide. The faults were lead is found are of the late Caledonian period, late Silurian to early Devonian with mineralisation taking place in the late Carboniferous or early Permian age. Permian is derived from Perm, a town in Russia.

Gold is found mainly on and around Clogau mountain above Bontddu on the Barmouth to Dolgellau road as well as deposits above the A470 between Dolgellau and Ganllwyd. There are 6 lodes associated with Clogau gold mine. The richest being called St. David's. Clogau was by far the richest but the main phase of extraction ended in 1911. It has been intermittently worked since then. The 'officially' recorded output of these mines was 165,031 tons of gold ore that made 78,507 ounces of pure gold.

Manganese outcrops in the Rhinogydd in several places. It was mined from around 1887 until around 1925. There are also outcrops of manganese on the Llŷn Peninsula at Rhiw and Porth Ysgo. Workings in the Rhinogydd were mainly on the surface. In 1917 twenty two men were employed and the impressive cuttings are easily seen at Llyn Eiddew Mawr. Workings started to spread uphill towards Llyn Dywarchen on the side of Moel Ysgyfarnogod and in the area close to Llyn Du. A rough track ran from here and it thought that a lorry made the journey here, taking a whole day for the round trip from the station at Talsarnau! The descent from Llyn Du to Llyn Eiddew bach is steep and it is probable that pack horses or sledges made this part of the journey. The uses of manganese were varied and its main uses were in the chlorine industry to produce bleach; the steel industry and glass making.

It is important that if you want to visit these mines DO NOT venture into any of the adits or levels. Some of the mines mentioned are situated in high mountainous terrain. Please dress accordingly when visiting and be prepared for bad weather. All these mines have flooded passages, loose rock and hidden shafts. However, there are companies who take adventurous minded people into some of the mines mentioned herein such as Pandora, Bwlch y Plwm and Parc mines. Unfortunately there is only one metal mine open to the public as a show mine, Sygun. It is well worth a visit. This is situated just off the A498, north of, but close to Beddgelert.

To locate the mines described the use of a map is important and with, obviously, the skill to read a one. A grid reference is given for each mine described along with the relevant map on which it can be located. The maps to use are: Ordnance Survey 1:25,000 OL17 Snowdon/Yr Wyddfa and Ordnance Survey 1:25,000 OL18 Harlech, Porthmadog & Bala/Y Bala. These give a more detailed picture than the general OS 1:50,000 Landranger maps.

1. ABERLLYN LEAD AND ZINC MINE and COED MAWR POOL LEAD MINE
2. ABERGLASLYN and BRYNFELIN COPPER MINES
3. BRAICH YR OEN and HAFOD Y LLAN COPPER MINES
4. BRITANNIA or SNOWDON COPPER MINE
5. BWLCH Y PLWM LEAD MINE
6. CEFN COCH GOLD MINE and the associated BERTHLWYD and NEW CALIFORNIA GOLD MINES
7. CLOGAU GOLD MINE
8. CRIB DDU or LLWYNDU COPPER MINE
9. CWM BYCHAN COPPER MINE
10. CWM CIPRWTH
11. CWM DWYFOR COPPER MINE
12. CYFFTY LEAD MINE
13. DRWS-Y-COED and associated mines SIMDDE DYLLUAN and BENALLT
14. GARTH GELL GOLD MINE
15. GLASDIR COPPER MINE
16. GWYNFYNYDD GOLD MINE
17. HAFNA LEAD MINE
18. HAFOD Y PORTH COPPER MINE
19. KLONDYKE LEAD MINE
20. LLANBERIS COPPER MINE
21. LLANRWST LEAD MINE
22. LLIWEDD BACH COPPER MINE
23. LLYN BODGYNYDD
24. LLYN DU MANGANESE MINE
25. LLYN DYWARCHEN MANGANESE MINE
26. LLYN EIDDEW MAWR MANGANESE MINE
27. PANDORA LEAD MINE
28. PARC LEAD MINE including VALE OF CONWY LEAD MINE
39. SYGUN COPPER MINE
30. YSTRAD/SILURIAN or GARREG FAWR IRON MINE

Snowdonia Metal Mines

Aberllyn Lead and Zinc Mine and Coed Mawr Pool Lead Mine

Map: Ordnance Survey 1:25,000 Explorer OL 17 Snowdon/Yr Wyddfa
Grid references: Aberllyn SH 794 579 and Coed Mawr Pool SH 779 584

It is uncertain when mining commenced at Aberllyn but it was first noted in 1826 whilst it is certainly possible that mining started before that date. It may well have been the first on the Estate. Mining was continued by a number of small partnerships with never more than 6 men between 1841 and 1878 when the first commercial proprietors arrived. During these early years lead was mined but in 1868 zinc started to be mined and eventually over a 6 year period it became the primary ore to be extracted.

Between 1869 and 1904 some 2,500 tons of zinc were extracted using a workforce in 1900 of over 200 men. The mill was a very substantial building having 7 floors stepped down the hillside. Final closure came in 1921.

The path up the gorge was used by the miners working at this mine and others further up the valley. More recently the path was used by the wives of the forest smallholders. These women struggled under heavy loads after going to the weekly market at Llanrwst.

Mining at Coed Mawr Pool has no records prior to 1838. The ruins before reaching Coed Mawr Pool are of Wheal George and given this name by the Wheal George Mining Company who worked the mine between 1880 and 1885. Unfortunately the investors in the mine were somewhat tricked by George Beckinsale, the owner of the company. Initially the mine was promoted as being '*one of more than fair promise*' and shares were offered at £1 a share. The promise of wealth came to nought and became a '*most wretched affair*'.

In September 1881 six men were

1. The remains of Aberllyn Mill; 2. Ruins of the works inside the mill

1. & 2. Part of the mill's exterior wall;
3. Entrance to Number 6 level

employed but by the end of December there were 12 which peaked at 46 by April the following year. The purchasers of the new shares were asked to pay £1. 10 shillings (£1.50) a share! No profit seems to have been made with mining ceasing in 1883. No subsequent interest was shown in the site.

Coed Mawr Pool is shrouded in total confusion with lodes given different names in different periods of time. Even

the names of shafts altered with each new company. It is claimed to have begun life in 1697. Sir John Wynn from Gwydir Castle is purported to have mined lead in the area close to Pencraig Pool and it is documented as a Wynn site. Mining ceased in 1930.

Both these sites can be visited by a great way-marked walk starting from the Conwy County Council Pont-y-pair fee paying car park in Betws-y-coed.

Aberglaslyn and Brynfelin Copper Mines

Map: Ordnance Survey 1:25,000 Explorer OL 17 Snowdon/Yr Wyddfa
Grid references: Aberglaslyn SH 594 462 and Brynfelin SH 589 472

Aberglaslyn can be reached from the National Trust fee paying car park at Nantmor. Walking over Pont Aberglaslyn a path leaves the A498, close to Aberglaslyn Hall, to the mine ruins. To the left of the ruins is a fenced off and very deep looking shaft whilst to right is an adit.

It is thought that the Romans knew that copper existed hereabouts. Aberglaslyn was worked from around 1769 and was purported to have good quality ore. It should be noted that at this time no road existed through the Pass but there was a port at Aberglaslyn from which the ore was shipped. In 1800 copper attracted a high price. Speculators became involved and produced a reasonable quantity of ore but mining was discontinued in 1819. Between 1804 and 1847 only 56 tons were

1. A shaft close to the barracks; 2. Barracks in the Aberglaslyn woodlands

Snowdonia Metal Mines

sold in Swansea. Mining seems to have ceased completely by the mid 19th century.

Nothing much remains of Brynfelin although a ruined building was possibly a barracks for the miners or the mine manager. The mine was certainly operating pre 1760 when it was said to be '*doing well*'. These were the most productive years. During the 19th century there were many failures and many owners. Ore was sold again in the 1830's and the owners of Brynfelin wanted to acquire Sygun copper mine in 1838. After another closure the mine re-opened again in 1851 and Brynfelin Copper Mining Company formed in April 1861. It appeared that there were 3 veins close together with the main lode yielding 7 tons of ore per cubic fathom (a fathom is 6 feet) and some 515 tons of ore were sold for £3,736. Due to many difficulties the mine closed very shortly after. It was offered for sale in 1864 but the new owners only produced 16 tons of ore in 1868 and once again by 1875 it had closed.

1. Looking down on Brynfelin copper mine with Yr Aran, Snowdon and Lliwedd in the background;
2. & 3. Brynfelin mine manager's house

Snowdonia Metal Mines

Braich yr Oen and Hafod y Llan Copper Mines

Map: Ordnance Survey 1:25,000 Explorer OL 17 Snowdon/Yr Wyddfa
Grid references: Braich yr Oen SH 614 516 and Hafod y Llan SH 623 523

Both these mines can be accessed from the Watkin Path up Snowdon that starts from the Snowdonia National Park fee paying car park at Bethania. Braich yr Oen is situated high above Cwm Llan just below a shoulder of Yr Aran 747 metres or 2,451 feet. There is an amazing vantage point by the mine. Snowdon 1,085 metres or 3,560 feet of course dominates the view but Y Lliwedd 898 metres or 2,946 feet. assumes a great presence too. Hafod y Llan is seen directly across the cwm and the spoil heaps and ruins of the Hafod y Llan slate quarry far below on the cwm floor add to the splendour. Having gained so much height it would be a pity to miss out on ascending Yr Aran and returning via Hafod y Llan slate quarry.

When walking up the Watkin Path and where it meets the Afon Cwm Llan are the remains of the Hafod y Llan works. These

1. Ruins of Hafan y Llan works; 2. Hafod y Llan mill by the side of the Afon Cwm Llan

Snowdonia Metal Mines

are easily spotted on the far side of the stream. Here were housed the main buildings. These included the mill which had waterwheels, roller crushers and dressing floors. From this there was once a bridge over the river towards Braich yr Oen. When walking up the incline towards Braich yr Oen stone blocks, on each side of it, have twin holes in each of them. These were used for brackets attaching the railway line and are, perhaps, the best example in Snowdonia. *On a note of history here Cwm Llan was the setting for the Khyber Pass in the comedy film 'Carry on up the Khyber'.* Where the incline meets the Hafod y Llan slate quarry tramway it is well worth turning left along it to view the amazing incline descending into Cwm Llan before continuing up to Braich yr Oen.

Braich yr Oen and Hafod y Llan often worked in conjunction with each other, so records are intertwined. Braich yr Oen is the older of the two and started life in 1762. It is recorded that copper ore was sold in Swansea in 1825/26 and again in 1840. Along with Hafod y Llan, Braich yr Oen supposedly produced 150 tons of copper ore and 30 tons of lead. In 1862 Bwlch yr Oen was re-started after a long break in production. After another break work commenced again in 1883 but the mine finally closed in 1886. The spectacular Hafod y Llan incline was built

1. Ruins at Braich yr Oen; 2. Ruins and spoil heaps at Braich yr Oen with the Watkin Path far below

Snowdonia Metal Mines

in the 1870's after the vision of a railway being constructed along the valley floor to Porthmadog. This never materialised and the ore was carried using horse power for the 17.6 kilometres or 11 miles journey to Porthmadog.

If an ascent of Yr Aran made with a return via Cwm Llan and the slate quarry, the Watkin Path descends past Gladstone Rock. On 13th September 1892 William Ewart Gladstone the Prime Minister of Britain for the 4th time delivered a speech from here to 2,000 people to officially

1. The ruins with Hafod y Llan slate mine in the background; 2. Braich yr Oen mine entrance; 3. Looking across to Hafan y Llan from Braich yr Oen; 4. Looking across Cwm Llan to the Braich yr Oen incline rising from the Watkin Path

open the Watkin Path. It was named in honour of Sir Edward Watkin a Liberal Member of Parliament and railway entrepreneur. He had retired to a chalet in Cwm Llan. In order for visitors to be able to walk up Snowdon Sir Edward created a

path from the already existing one up to the slate quarries to the summit of the mountain. It was the first designated footpath in Britain and the first step towards the opening up of the mountains and countryside for walkers. Gladstone was 83 at the time of his speech!

Continuing down Cwm Llan the path reaches a ruin on the left. This is Plas Cwm Llan. This used to be the home of the Hafod y Llan Slate Quarry Manager. During the Second World War soldiers used this building as a target when training for 'D Day'. The holes in the walls of the house are bullet holes.

Snowdonia Metal Mines

Britannia or Snowdon Copper Mine

Map: Ordnance Survey 1:25,000 Explorer OL 17 Snowdon/Yr Wyddfa
Grid reference: SH 616 547

Britannia copper mine opened in the early 1790's. It is situated immediately above Glaslyn. The miners barracks were erected shortly after. During a particularly bad winter the miners had to tunnel through snow to reach their work with drifts up to 18.2 metres or 60 feet deep in places! Ore was being sold for 15 years in the period between 1804 and 1842. In 1805 a tourist noted that a track from Snowdon summit to the Saracen's Head (now the Snowdon Ranger Youth Hostel) was being used by two horses to sledge the ore down. Men struggled to the top of Snowdon loaded with sacks of ore for the horses to cart it to down to the Saracen's Head From there the ore went by cart to Caernarfon for shipment. Once the road had been built from Llanberis over the Pass it was much easier to transport the ore on that. The Miners' Track was built to reach this at Pen y Pass and was in place by 1813.The mine was worked intermittently, although

1. The mine close to the shore of Glaslyn with the ruined barracks below the mine; 2, 3. & 4. The ruins of the Britannia Mine crushing mill

pretty unsuccessfully, for over 100 years by a number of companies until closure in 1916. There were 8 levels some of which remain along with the ruined mill and other associated buildings close to Llyn Llydaw.

The causeway across Llyn Llydaw was built in 1853 to serve the Britannia copper mine. Before this was built a raft was used to carry horses and wagons full of copper ore across the lake. The water level was lowered by 12 feet to enable the causeway

to be built. During the building a prehistoric oak dugout canoe 10 feet or 3.04 metres by 2 feet or 61 centimetres was found. This was indeed proof that humans wandered around here for thousands of years before us.

On an historical note the low point, below Snowdon 1,085 metres or 3,560 feet and Lliwedd 898 metres or 2,946 feet, is known as Bwlch y Saethau, *Pass of the Arrows*. Miners coming up from Beddgelert used to come over this pass using fixed iron chains!

King Arthur is presumed to have fought his last battle here although many other places have laid claim to this. His death was supposed to be during the fight against the residents of a fort on Braich Tregalan in Cwm Tregalan marked on the OS map. His enemy were being forced uphill but on reaching the col and realising escape was impossible a hail of arrows was shot, one of which killed Arthur. On two occasions Bedivere tried to throw his sword into Glaslyn to no avail. On the third attempt a hand rose up from the depths grasped the sword then disappeared with it into the lake. The story continues when Arthur was carried to the lake shore a small boat appeared with three maidens on board dressed in white. They carried him off to the island of Afallon for his wounds to heal.

His knights are supposed to have gone to sleep in a cave on Lliwedd. The cave can be found on a climb called Slanting Gully. Here they wait until such time as they are needed by Wales. There is also a cave on Dinas Rock, close to Pontneddfechan in South Wales where Arthur's knights are supposed to be sleeping!

The mine ruins are easily reached by following the Miners' Track from Pen y Pass. Crossing the causeway the ruins of the crushing mill are quickly reached. Continuing past this leads up to Glaslyn and the mine.

Snowdonia Metal Mines

Bwlch y Plwm Lead Mine

Map: Ordnance Survey 1:25,000 Explorer OL 18 Harlech, Porthmadog & Bala/Y Bala
Grid reference: SH 626 413

It is possible that the Romans were the first to mine here. In an 1876 survey of the mine one of the levels was noted as *The Roman Lode*. The earliest records date back to May 1726. There were several owners who made very little if anything at all. In 1875 the Penrhyndeudraeth Mining Company dug a deep level adit to drain the mine working and to cut across the lode of lead. A new mill was built close to the site of the farm lower down the hillside. In the same year the company was wound up having sold less than a hundredweight of ore if any at all!

There was talk at one time of combining the workings of Bwlch y Plwm and the Jane and Catherine Consols 731 metres or 800 yards to the south east. Work still continued after Bwlch y Plwm was wound up but foundered in 1877. Alas the fine engine house chimney at Jane and Catherine Consols was demolished by the

1. One of the open workins; 2. Ore bin

Forestry Commission in 1965 during their period of 'removing the scars of industry phase'. There is little to see there as there is much thick forest.

The mine can easily be reached from Rhyd but there are many open shafts and care is needed.

Cefn Coch Gold Mine and the associated Berthlwyd and New California Gold Mines

Map: Ordnance Survey 1:25,000 Explorer OL 18 Harlech, Porthmadog & Bala/Y Bala Grid references: SH 717 234 and Berthlwyd SH 722 235

The gold mines of Cefn Coed/Berthlwyd Mines are part of the Dolgellau gold ore belt. Although a much smaller concern than either Clogau or Gwynfynydd Mines it was the fourth richest in the area. Cefn Coch was mined in conjunction with Berthlwyd Mine and the history of the two mines are intertwined and difficult to distinguish between.

The first mining took place at Berthlwyd in 1845 but it was not until 1862 that gold was discovered at Cefn Coch. Prior to that date there are no figures to show how much gold had been produced although, reportedly, much ore was mined. Between 1862 and 1865 the gold yield was 648 ounces which was obtained from 1,900 tons of ore! The mines closed in 1866. An attempt to re-work the mine again between 1873 and 1877 was unsuccessful.

In 1887 a 'Gold Rush' occurred in the area. On the back of that Cefn Coch re-opened in 1889. It's most successful year was in 1894 when it yielded 180 ounces of gold from 360 tons of ore. The total of gold yielded during 1862 and 1912 was 1,392 ounces (87 pounds). Final closure came in 1914.

Cefn Coch can be reached from Ganllwyd by initially following a lane, then a path to the superb Rhaeadr Ddu. A way-marked path from the waterfalls leads to the mine. On the way to the waterfalls there is a unique inscribed slate tablet carved with a romantic story. It took much painstaking effort to translate the writing as some of the words were in Latin. The problem was solved by Eggerton Philimare. He had discovered, in a copy of the antiquarian 'Bygones', that the lines were from Thomas Grey's Latin 'Alcaic Ode written at the Grande Chartreuse' in 1740

1. The ruined mill area; 2. The settling ponds looking towards the barracks and tramway

Snowdonia Metal Mines

whilst he was travelling in Europe with his friend Hugh Walpole. The National Trust eventually translated the relevant verse:
On the face of the adjoining rock, in the late 18th century, lines from the alcaic ode by Thomas Grey were cut by an unknown hand, the inscription was in latin here accompanied by an English Tranalation.

'O, thou! the Spirit 'mid these scenes abiding,
What'er the name by which thy power be known
(Truly no mean divinity presiding
These native streams, these ancient forests own
And here on pathless rock or mountain height
Amid the torrent's everxechoing roar,
The headlong cliff, the wood's eternal night
We feel the Godhead's awful presence more
Than if resplendent neath the cedar beam,
By Phidias wrought, his golden image rose),
If meet the homage of thy vot'ry seem
Grant to my youth – my wearied youth –
 repose'.

However, the above unknown hand could well have been that of William Alexander Madocks (1774–1828). He gained much of his fame by building the Cob at Porthmadog. The tablet could also have been carved by one of his long staying visiting poets, including Shelley. Madocks inherited a modest fortune and spent some of it purchasing the small Dolmelynllyn Estate. The larger part of this inheritance, though, was held in trust.

1. The tramway and mine buildings;
2. The tramway leading away from the main buildings towards a mine entrance;
3. One of the mine entrances

Snowdonia Metal Mines

Clogau Gold Mine

Map: Ordnance Survey 1:25,000 Explorer OL 18 Harlech, Porthmadog & Bala/Y Bala Grid references: Old Clogau SH 678 198; St David's lode SH 676 202; Clogau No 2 SH 673 201; Lefel Fawr, Ty'n y Cornel SH 672 201; Llechfraith Adit SH 668 191

There are 6 gold bearing lodes on Clogau mountain with the most rich being the St. David's lode where half the regions gold has been found. Originally a copper mine it was only by chance that gold was discovered. Copper mining ceased in 1860. It is thought that the Romans mined here but only for copper and possibly lead. Between 1825 and 1842 both the Old Clogau and Figra Mine on the opposite sides of the gorge were mined for copper. The chance discovery of gold came in 1854. This lead to the start of several 'gold rushes' in the area. Having found gold the mine simply became known as Clogau, with Clogau 2 being the most productive.

During peak production 63 men worked above ground with around 190 below. Between 1862 and 1911 165,031 tons of gold ore produced 78,507 troy ounces of gold. The peak year was in 1904 when 18,417 troy ounces of gold were produced. The main phase of mining ended in 1911 but intermittent mining has continued since then but now there are plans to re-open the mine.

Llechfraith Adit started being dug in 1880 and by the end of 1881 it had reached 336 metres or 1,102 feet long. A new company extended it further between 1889/90 but with little return for their efforts and expenditure. However, gold was eventually found and a new company formed from Barmouth businessmen who then capitalised on this expenditure by employing 25 men between 1891 to 1898. The gold ore was transported from the adit for crushing in 6 cwt buckets on an 1,006 metres or 1,100 yards long aerial ropeway that descended 64 metres or 210 feet to Figra Mill. There are ruins hereabouts and spoil heaps.

Many of the mine remains can be seen during a circular walk from Bontddu.

1. Tramway and Y Garn; 2. Part of the old tramway above a lower entrance into Clogau; 3. Ty'n y Cornel adit into Clogau

Crib Ddu or Llwyndu Copper Mine

Map: Ordnance Survey 1:25,000 Explorer OL 17 Snowdon/Yr Wyddfa
Grid reference: Crib Ddu/Llwyndu SH 606 482

Crib Ddu is usually known as Llwyndu although the hill the mine is situated is Crib Ddu 318 metres or 1,043 feet. The mine is situated on the opposite side of the hill to Sygun. Walking up the steep hillside from Beddgelert is the start of a grand circular walk that also takes in Cwm Bychan copper mine and returns to Beddgelert by the Fisherman's Path by the side of the Afon Glaslyn.

Sygun Mine was not very profitable so attention was turned to Llwyndu but like Sygun things went from bad to worse. Llwyndu was a virgin find and in a letter dated 25th July 1838 it was said that the percentage of ore was 15%. In 1839 ore was found that was almost 30%. Interestingly the dressing was done by 20 girls and it was a godsend for the mine as they said it was "*the cheapest thing we have on the mine*". It appears that without them the mine could well have folded before it did in 1844.

The area has well preserved remains. These include the mine processing site with the remains of some structures such as the processing floor, the mine managers house or office, spoil heaps, tanks and a 'calcining' flue for extracting arsenic! Calcining is a method that reduces, oxidises or desiccates (removing all moisture) by roasting or exposing to heat. Nearby to the north east is the shaft from which the extracted was raised to the surface. Llwyndu is associated with Sygun.

Beddgelert is well worth exploring as it very pretty. The legend of Gelert is well documented and even lends itself to a book. Briefly the legend is as follows: *Some 800 years ago Prince Llywelyn set out on a hunting trip with his hounds and huntsman. His baby son and heir, was left asleep in his cradle at home. The servant girl looking after him was far from reliable. As soon as the*

1. & 2. The mine at Bwlch y Sygun;
3., 4. & 5. Ruins below the mine

hunting party had left she went off to meet her lover for a stroll along the river. Llywelyn had a favourite hound called Gelert and he was puzzled as to why he could not see him at the hunt. He presumed the hound had gone back home but a premonition that something dreadful had happened sent Llywelyn galloping home. On arriving he found the floor awash with blood and torn bedclothes.

Snowdonia Metal Mines

The cradle was empty and overturned. Gelert appeared with blood dripping from his mouth. Unfortunately for Gelert, Llywelyn acted on impulse and presuming the dog had killed his son he drew his sword in a rage and killed the hound. The dying howl from Gelert had a feeble echo, seemingly coming from under the bedclothes. Dragging them aside he found his son underneath safe and sound. The body of a huge wolf was by his side, dead. Gelert's instinct had made him return home in time to save the life of his master's son. Frantic with sorrow it is said that Llywelyn never smiled again.

Just after starting the walk and before the open hillside it passes the home of Alfred Beatall M.B.E. he is famous for illustrating the Rupert Bear stories: Rupert

Bear was created by the English artist Mary Tourtel. The first comic strip first appeared in the *Daily Express* on 8 November 1920. In 1935 the Rupert stories were taken over by Alfred Edmeades Bestall M.B.E. an artist and storyteller. Born in Mandalay, Burma on 14th December 1892, he died on the 15th January 1986. Alfred lived in the house from 1956 to 1986. His first story was published on the 28th June 1935 and the last on the 22nd July 1965 although he still did covers for Rupert Annuals until 1973.

The character Rupert Bear lives with his parents in a house in Nutwood, a fictional idyllic English village. He is depicted wearing a red sweater and bright yellow checked trousers, with matching yellow scarf. Usually seen as a white bear he was originally brown and was made white to save on printing costs.

The majority of the other characters in the series are also anthropomorphic animals (animals with humanoid forms). Regardless of species they are all drawn roughly the same size as Rupert referring to them as his "chums" or "pals". His best friend was Bill Badger. Others were an elephant (Edward Trunk), a mouse (Willie), Pong-Ping the Pekingese, Algy Pug, Podgy Pig, Bingo the Brainy Pup, Freddie and Ferdy Fox, and finally Ming the dragon.

Rupert was helped on many of his adventures by the kindly Wise Old Goat who also lives in Nutwood. The few main human characters in the stories were the Professor (who lives in a castle with his servant), Tiger Lily (a Chinese girl), and her father 'The Conjuror'. Perhaps Alfred's most famous drawing was 'The Frog's Chorus'. This inspired the cartoon video 'The Frog Song' composed by Sir Paul McCartney.

Looking up Cwm Bychan to Crib Ddu

Cwm Bychan Copper Mine

Map: Ordnance Survey 1:25,000 Explorer OL 17 Snowdon/Yr Wyddfa
Grid reference: SH 602 472

The first mining here was in 1720. The heydays were between 1782 and 1802. In the early 1870's the mine was run by the Cwm Buchan Lead Mining Company but ceased trading by 1875. A brief flurry of activity took place in the 1920's even to the extent of erecting a1.4 kilometres long aerial ropeway. It was constructed to convey ore from the workings at the top of the cwm to the mill at the bottom close to the Welsh Highland Railway. The aerial ropeway conveyed chalcopyrite, a crystalline mineral, the principal copper ore. It has a brassy yellow colour.

The mill area, just beyond the railway bridge above the car park, has two circular buddles on the left. These were part of the dressing plant for the mines. There are also other remains including sundry concrete bases and a curious metal cage made of iron bars with a pulley at one end. As previously mentioned in the description to Crib Ddu/Llwyndu, Cwm Bychan can be included. However, if the steep ascent above Beddgelert does not appeal a shorter, but still uphill, walk can be from the National Trust fee paying car park at Nantmor close to Pont Glaslyn.

1. Buddle at Nantmor; 2. The lower mine area in Cwm Bychan; 3. One of the aerial ropeway support towers with mine tailings to the left; 4. A collapsed aerial ropeway tower with mine tailings; 5. Curious metal cage at Nantmor

Cwm Ciprwth

Map: Ordnance Survey 1:25,000 Explorer OL 254 Lleyn Peninsula East/Pen Llŷn Ardal Ddwyreiniol
Grid reference: SH 525 477

Associated with this mine is one called **GILFACH** found at grid reference SH 531 477. The adit into this is 305 metres or 1,000 feet long is located at the bottom of the hill just after starting the walk up to Cwm Ciprwth. Although only 30 minutes from the road Cwm Ciprwth is a remote site.

Ore was being produced certainly as early as 1828 when 40 tons of ore was sold. In 1890 there were 12 men working below ground and another six above. By 1891/2 half that number were employed. The mine closed and went into liquidation in 1894.

The 7.26 metres or 25 feet diameter water wheel is a superb feature and was made by Dingey and Sons of Truro. It worked the pumps and winding drum by using a very basic friction clutch. A wooden angle bob is above the flooded engine shaft. This supported the pump rods and transmitted motion from a crank on the waterwheel axle via a line of iron and wooden flat rods. There are a couple of stone buildings that may have been used as a smithy, stores and miners barracks. The site has been renovated quite recently. The original installation dates back to 1889/90. It is a unique survivor of water power that was employed for pumping and winding without noise or pollution.

A way-marked path leads to this fine site from Cwm Pennant from close to the gated modern bridge over the Afon Dwyfor. It is well worth the effort to see the restored waterwheel if nothing else.

1. General view of the mine area; 2. The waterwheel; 3. The entrance into Gilfach Mine

Cwm Dwyfor Copper Mine

Map: Ordnance Survey 1:25,000 Explorer OL 254 Lleyn Peninsula East/Pen Llŷn Ardal Ddwyreiniol
Grid reference: SH 541 506

Known at one time as Blaen y Pennant it was replaced by the name now usually used. The mining Company of Wales took over the running of the mine in 1850 but it was certainly mined prior to then. There were lodes of lead and copper and was worked to a depth of 7 fathoms (42 feet) but work was suspended due to the lack of water and machinery. It was subsequently mined in 1868 by the Cwmdwyfor Copper and Silver-Lead Mining Co. At this time the .91 metre or 3 feet gauge Gorseddau Tramway, constructed in 1856, was operating. This was servicing the Gorseddau slate quarry taking slate to the mill and thence on to Porthmadog. In 1872 a Parliamentary Act authorised the conversion of this line to 1 foot 11½ inches, with a connection from Cwmystradllyn up to the Prince of Wales slate quarry mill. In 1876 the loop to Cwm Dwyfor was working. The mine was liquidated at the end of 1876 but resurrected in April 1877. Results continued to be very disappointing and mining ceased, forcing closing in 1878. The mine was offered for auction again in 1879.

Cwm Dwyfor had the enviable distinction of being the only copper mine to be directly connected by a tramway to the port in Snowdonia.

There is much to see at this interesting site. The principal remains include the wheel-pit which housed the 10.6 metres or 35 feet by 1.22 metres or 4 feet waterwheel. The pit itself was an astonishing 12.5 metres or 41 feet by 5 feet 9 inches by 1.75 metres lined with massive stone blocks. Most of the building remains are composed of large stone blocks whilst the crusher house is also very solidly built with buttressing on its south side. Beyond the main area there a row of terraced one story cottages that were probably used by the miners.

1. Looking up the Cwm Dwyfor incline;
2. Looking down Cwm Dwyfor incline;
3. Cwm Dwyfor copper mine

The walk to this mine commences from the road end in Cwm Pennant. Venturing into high ground the path is way-marked to start to where the path splits. It then contours around on the intermittently seen tramway to the incline. Walking up this leads into Cwm Dwyfor.

1. The location of Cwm Dwyfor showing the incline just right in the bottom half of photo; 2. The remains of the wheel-pit; 3. & 4. General view of the Cwm Dwyfor copper mine; 5. The Barracks

Cyffty Lead Mine

Map: Ordnance Survey 1:25,000 Explorer OL 17 Snowdon/Yr Wyddfa
Grid reference: SH 772 588

Little is known of Cyffty mine and like so many others in the area it seemed to have been operational since 1787. Known as Pencraig in 1846 it was re-named *Bettws y Coed Mine* in 1878 and kept this name until 1899 when it became Cyffty Mine. Mining continued until the company went into liquidation owing £5,000 in 1908. A single caretaker stayed at the mine until after the outbreak of the First World War. In 1915 a new company worked the mine until 1917 when it became a public company. Employment increased to a maximum of 30 men from 1918 to 1921. Another new company took over in 1925 and after several other owners the mine finally closed in 1958 at the same time as Llanrwst Mine.

The engine house at Cyffty with its fine chimney was wantonly destroyed in 1966.

This mine can be approached from the Llyn Sarnau car park which when translated means '*the lake of old tracks*',

1. The gated Western Pumping Shaft;
2. Ruins of works by the side of a wheel pit;
3. The Crusher House ruins below the parking area

aptly named as there are many tracks hereabouts. A circular walk also taking in Llanrwst lead mine is easily followed from the car park.

Snowdonia Metal Mines

Drws y Coed and associated mines Simdde Dylluan and Benallt

Map: Ordnance Survey 1:25,000 Explorer OL 17 Snowdon/Yr Wyddfa
Grid references: Drws y Coed SH 546 534, this reference is for the ruined cottages, although other mine remains are east and west of this with a deep shaft above the spoil heap above the turning to Tal y Mignedd camping and caravan site; Simdde Dylluan also known as Owl's Chimney or Talysarn SH 541 534. On the opposite side of the valley is Benallt SH 536 635.

The remnants of Drwysycoed and Simdde Dylluan are easily seen a few yards beyond the roadside wall on the B4418 Penygroes to Rhyd Ddu road. There are many enterable workings. However, all these mines have long been abandoned and are in a VERY DANGEROUS condition and entry is most ill advised. There is a remarkable lack of spoil at this site.

Apparently Edward the 1st visited the site in 1284. In 1756 mining took place but in 1761 at Simdde Dylluan there was more success as was the period at the end of the 18th century and early 19th. In this latter period there was a huge demand for copper during the Napoleonic Wars. Between 1821–1840 the mine realised its potential and produced 6,000 tons of ore

Snowdonia Metal Mines

that was sold in Swansea. Before the Nantlle Railway opened in 1828 ore was transported from the mine using mules or ponies on the old Roman road to Caernarfon. Starting in Beddgelert the road went past Rhyd Ddu before turning west on the slopes of Mynydd Mawr and on to Fron down to Bodaden before crossing Y Foryd to Dinas Dinlle. In all, between 1804 and 1913, the mine realised some 12,939 tons of ore. This figure may well be an underestimate and probably double the amount of ore was produced.

The Nantlle Railway was a horse powered 3 feet 6 inches gauge tramway. In the first 6 months of 1831 the mine produced 522 tons of ore of which 100 tons were shipped to St Helens on the 1st January. The whole of that year amounted to 629 tons but in 1832 it increased to 916 tons which was taken to Swansea.

Along the access road to Drws y Coed Isaf Farm, a part of the Snowdonia Slate Trail, is the remnant of a chimney like structure. This is in fact hollow and was used to bypass water to adjust the power output. In Welsh these structures were known as 'cafn gwyllt', *wild spouts* and is an interesting feature. They were also constructed from wood.

Below the access track to Tal y Mignedd camping and caravan site some 250 metres from its junction with the B4418 are the remains of other structures in particular a fine wheel pit. It can be seen from the road.

1. & 2. An entrance into the mine;
3. The miners cottages at Drws y Coed;
4. General view of ruins at Drws y Coed

Snowdonia Metal Mines

Garth Gell Gold Mine

Map: Ordnance Survey 1:25,000 Explorer OL 18 Harlech, Porthmadog & Bala/Y Bala
Grid reference: SH 688 200

Garth Gell was a small and unsuccessful mine and is one of around 300 gold mines in the gold mining area! Several short adits were driven, including two on the eastward extension of the Clogau, St David's lode but mineralisation was poor. The only recorded output was in 1900 when 5 ounces of gold were produced from 26 tons of ore. The mine closed in 1902. Other minerals found here included chalcopyrite, pyrite, sphalerite and galena.

Welsh gold has always fascinated people and is the rarest precious metal in the world. It is lauded in songs of the ancient Bards taking on magical properties. Prior to 75 A.D. panning for gold was undertaken in the alluvial river beds.

The walk to this mine starts from the Fiddler's Elbow car parking area on the A496.

1. The Smithy; 2. & 4. Buddles; 3. Wheel-pit

Snowdonia Metal Mines

Snowdonia Metal Mines   47

Glasdir Copper Mine

Map: Ordnance Survey 1:25,000 Explorer OL 18 Harlech, Porthmadog & Bala/Y Bala Grid reference: SH 739 225

Glasdir mine was one of the most extensive mines in the Dolgellau area. It was first opened around 1852 as a quarry. Between 1872 and 1913 some 13,077 tons of dressed copper left the mine. In 1896 the mine was bought by the brothers Frank and Stanley Elmore along with their father William. By 1907 they had developed the Elmore Flotation Process. They had previously used the process at the Sygun Copper Mine but when that mine stopped working he transferred all the equipment to Glasdir. The mine finally closed in 1915. In terms of precious metals the mine produced 8,275 ounces of silver and 735 ounces of gold between 1872 and 1915.

The flotation process entails using the waste material and water. This waste usually has some 1% – 2% of copper ore and is crushed. Water is then added and fed into a ball mill which is a rotating drum with steel balls that crush the ore into a fine powder to form slurry. This is then fed into a rake classifier when it comes out of the mill and then into flotation cells. Any particles that are too large are returned to the mill. Air is injected into the flotation cells where foaming agents are added. This creates froth and copper particles, because of their light weight, become a part of this froth. Heavier particles such as iron sink. The mixture of copper and froth containing around 20% – 40% copper continues for further processing to extract the copper itself.

A short circular walk can be followed from the Glasdir Natural Resources Wales car park to view the workings. There are a number of bays in which to park and an information panel is at the far end close to the road junction.

The Glasdir Mill remains

Gwynfynydd Gold Mine

Map: Ordnance Survey 1:25,00 Explorer OL 18 Harlech, Porthmadog & Bala/Y Bala Grid reference: SH 735 282

Although gold was discovered in 1863 it was not until 1887 that any real profit was made. It was a laborious job with a working day both hard and long. In 1888 Hugh Pugh from Dolgellau started work here. He wrote diaries which gave a fascinating insight into the methods and conditions prevailing at that time. Some 200 men were employed with many of them boys under 15. They worked a 10 hour shift!

Many of the miners lived in barracks at the mine although some took up lodgings in nearby farms whilst others had long walks from their homes each day whatever the weather. The 'white collar' people in charge lodged away from the miners in a separate boarding house. Hugh for example walked 8½ miles. He told of meeting at a bridge with others at 03.00 to start work at 07.00 with their 'wallets'. In one end would be a big homemade loaf whilst at the other would be rations.

For those staying at the mine a horse and cart travelled to Dolgellau each day for supplies. Once a week bread and other groceries were delivered to the mine. However, the menu on offer seemed somewhat monotonous. One miner reflected his view as follows –

'Rabbits young, rabbits old,
Rabbits hot, rabbits cold,
Rabbits tender, rabbits tough,
Thank the Lord, we had enough'

There was no entertainment although sing songs were usually held on Friday nights with a concert once a month. One of the mine managers opened a reading room and started a Bible class. Wages were poor and in 1900 a miner was paid 3 shillings and 6 pence (19 new pence) and 1 shilling and 6 pence (8 new pence) for a boy whilst a poor donkey only received 1 shilling and a penny (6 pence)!

The Dolgellau Gold Belt heralded the 'Welsh Gold Rush' in 1860 with many

The main ruins with Rhaeadr Mawddach beyond

companies prospecting in the area. The Ty'n Groes Hotel (then known as Thornton's Hotel) became the focal and meeting place for the miners. The gold belt extends for about 20 miles and 20 mines were operational at the peak of production but many were tiny affairs with few miners. The rock dates back to the Cambrian period some 550 million years ago

The first 'rush' was over by 1865 and for the next 20 years or so intermittent concerns continued until 1887 when William Pritchard Morgan owned the mine. He was eventually dubbed the 'Welsh Gold King'. Reputedly he made two fortunes here. The new discoveries were better organised, better equipped and with better finance. During this period 30 mines became operational but again many were small undertakings being little more than mere scratchings on the surface. Pritchard discovered a lode that became known as the Chidlaw Lode. It was one of several but this was the richest. Gold ore is considered rich if it contains 1 ounce of gold per ton of rock! 8,745 ounces of gold were produced from 3,844 tons of rock. This was more than double the amount of richness with 2.27 ounces of gold per ton. Pritchard once boasted that in two weeks he produced 2 stone more gold than his weight! In 1894 Pritchard discovered another lode that produced 10,000 ounces over a 2 year period. He sold the mine in 1900 but bought it back again in 1913 but with limited success this time and finally the mine closed in 1916. By comparison the amount of gold produced here between 1863 and 1916 would amount to £20 million.

Traditionally gold produced in Wales was used for wedding rings for the Royal Family. The ingot of gold presented to Queen Victoria was exhausted by 1960. In 1986 a kilogram ingot of 99% pure gold was presented to Queen Elizabeth to commemorate her 60th birthday and a smaller one given to the Duke of York.

Gold was extracted from the ore in the Gwynfynydd Mills. These were powered by a waterwheel and turbines. The ore was tipped onto screens then fed into stonebreakers before being crushed by stamp mills then being washed over copper plates covered with amalgam that picked up the gold. Grinding Pans, known as 'Britten Pans' were used to recover the visible gold from the higher grade ore. Smelting was also carried out here. Hugh Pugh who was mentioned above became a foreman and in one of his diaries he records that the Britten Pans worked from

Sunday midnight through to Saturday midnight and operated by two men working 12 hour shifts!

A circular walk taking in the mine area can be started from the Tyddyn Gwladys car park in Coed y Brenin. Not only does it take in the mine it has superb views of two fine waterfalls, Pistyll Cain and Rhaeadr Mawddach.

1. Forest enclosed spoil tips at Gwynfynydd Gold Mine (Photo: Eric Jones, Wikimedia Commons); 2. Gwynfynydd Gold Mine today (Photo: Malcolm Street, Wikimedia Commons)

Hafna Lead Mine

Map: Ordnance Survey 1:25,000 Explorer OL 17 Snowdon/Yr Wyddfa
Grid reference: SH 780 602

Mining commenced at Hafna around 1615 by Sir John Wynn. After analysing samples he found the ore was of sufficiently high grade to commence mining operations. Mining continued not only here but at 21 other sites in the Gwydir forest. In 1819 Hafna was being worked by the landowner of the Plas yn Cefn Estate, Edward Lloyd. The first mill was constructed in 1879.

The machinery was initially operated by a waterwheel and subsequently by a gas engine. The main area of the mill occupied four floors where the ore was crushed, sorted and concentrated prior to smelting. Ore was brought into the mill area via an adit and shaft. The adit is now locked but can be seen on the left at the bottom of the long flight of steps near the top of the mill area. The shaft is also securely covered.

The ore was stored in bins and sorted by hand on the topmost floor, floor one. From there it was mechanically crushed on the next floor down, floor two. This was then 'jigged', a process involving a wooden box 1.5 metres or 5 feet or more in length inside of which was a sieve. It was

1. Looking up the ruins of Hafna mine works from the car park; 2. Looking down the ruins of Hafna mine works from the midway point on the steps; 3. The only open tramway tunnel

suspended from above by a large pole. The box was three quarters filled with water and the crude ore placed on the sieve and jerked up and down in the water by means of the pole. Apparently the process had been invented in Derbyshire in the early 19th century.

The larger pieces of ore were removed for smelting. What was left was then 'buddled'. A buddle is basically a shallow circular construction where small sized ore was cleaned of its dirt by running a stream of water through. This method had been used elsewhere since the mid 1500's.

All the waste was taken away from all the milling stages through the tram tunnels, only one of which goes all the way through, to be dumped. The fine residues from the mill were collected in the slime pit and occasionally re-processed.

The Hafna smelting mill is unique in the Gwydir forest and was built in the 1880's. The smelting process effectively roasted the ore. As the temperature increased the molten metal would flow to the bottom of the furnace from where it was tapped. The fumes from the smelting process were toxic so a long flue carried them away and up the chimney well above the site.

A pleasant scenic way-marked walk starting from the Hafna Mine car park takes in much of everything about Hafna and Parc lead mines.

Snowdonia Metal Mines

Hafod y Porth Copper Mine

Map: Ordnance Survey 1:25,000 Explorer OL 17 Snowdon/Yr Wyddfa
Grid reference: SH 780 602

Hafod y Porth mine dates back to 1755 and was worked for its copper until around 1890 when it slid gradually into oblivion. When the mine came up for sale in 1845 there was a 6.7 metres x .91 metres or 22 feet x 3 feet waterwheel, .6 metre or 2 feet rollers and an 8 head stamp and some 10 tons of 'T' section rail. In 1864 the mine was taken over again by the Hafod y Porth Copper mining Company with some ore being sold in Swansea. Profits gradually increased gradually until 1868/69 but by 1873 they had once again dwindled. Subsequently a succession of companies tried and failed. Between 1882 and 1884 the mine was called Maudsley, after the then owner. In 1888 it became known simply as Aran. Other owners came and went with some probably never even putting a spade into the ground!

The levels and shafts are home to a variety of bats. Choughs also breed in the area and are known locally as 'y fran goesgoch', *'red legged crows'*. Choughs are members of the crow family along with Rooks, Crows and of course Ravens, the largest birds of the family.

Note the copper sculpture beyond the wall to the left. This was built to commemorate the women who worked for the copper mines.

A partially way-marked circular walk from Craflwyn takes in the sculpture and the mine. It is very scenic but there are some large unprotected drops in the mine area. They have loose or slippery edges and great care is needed. DO NOT climb over any barriers or approach the entrances. There is much to see without venturing close to these.

1. Ruins of the barracks; 2. The circular ruin close to the barracks; 3. Ruins; 4. Arch under a dressing floor

Snowdonia Metal Mines

Klondyke Lead Mine

Ordnance Survey 1:25,000 Explorer OL 17
Snowdon/Yr Wyddfa
Grid reference: SH 765 621

Klondyke Mill was a processing mill and constructed in 1900. It was built to process lead and zinc ores from Pandora Mine some 2 miles distant. In actual fact the mill was little used as Pandora Mine was not profitable after the mill was built. Pandora closed in 1905 and the mill itself in 1911. One of the main claims to fame came about in the 1920's when it became part of a notorious scam. Investors had been sought with the promise of much silver to be found and were basically duped. It was because of this is how the Mill became to be known. When the Mill did operate it was known as Geirionydd Mill and then as the New Pandora Lead Works. Today it is the largest building associated with lead mining in North Wales and is registered as an ancient monument under the guardianship of CADW and the only mine structure in the Gwydir forest to be so.

Klondyke Mine was basically a single adit and the portal is easily seen from the mill on the far side of the stream and directly below the aerial ropeway/incline seen earlier. Inside three tunnels go right and were driven a short way with a view to extending the mine.

A grand walk to this mine is best started in Trefriw and takes a high level approach to Llyn Geirionydd before descending to the mill.

1 & 4. The Klondyke lead mine ruins at SH 76497 62155 (Photo: Terry Hughes, Wikimedia Commons); 2. Aerial ropeway terminus at Klondyke (Photo: Malcolm Street, Wikimedia Commons); 3. The slag heap remains of the Klondyke lead mine (Photo: Terry Hughes, Wikimedia Commons)

Llanberis Copper Mine

Ordnance Survey 1:25,000 Explorer OL 17
Snowdon/Yr Wyddfa
Grid reference: SH 597 587

This mine produced much ore but made very little if any profit! The workings can be seen very easily from a large lay-by (part of the old road) on the A4086 road at the Nant Peris end of Llyn Peris. The workings, although not very wide, stretch up the hillside and are marked by several spoil heaps. Higher up there are also the remains of dressing floors and small stone buildings as well as extensive and dangerous open-cut workings In the days before the road was constructed all the raw material had to be taken out by boat down the lake and to bring heavy mining equipment in. There was only a path used by horses between Nant Peris and Llanberis. Mining commenced around 1760 and continued intermittently. The concern was officially dissolved in 1885 but production had certainly ceased well before that.

The tonnage of ore produced was in the order of 7,499 during its period of operation. In 1832 an astonishing 1,169 tons were produced, a record for a Caernarfonshire copper mine. In 1870 an astonishing amount of high quality ore yielding 34% of copper was produced. This was a figure that was, rarely, if ever, reached by any other UK copper mine.

At the time of writing it was not possible to visit this mine.

Snowdon viewed from Braich yr Oen. Hafod y Llan slate quarry at bottom of picture.

Llanrwst Lead Mine

Ordnance Survey 1:25,000 Explorer OL 17
Snowdon/Yr Wyddfa
Grid reference: SH 779 593

The Llanrwst engine house was built in 1876/77 to house a .64 metre or 25 inches cylinder horizontal condensing engine. This was mounted on the central block. Steam to work the engine came from an 11 ton Cornish boiler in the boiler house. The chimney is 18 metres or 59 feet high and was erected to provide a through draught for the fire in the boiler house. An elaborate pumping system was also operated by the engine, helped by a flywheel, to pump water out of Endean's Shaft to keep it dry. The engine house and its associated features are the ONLY surviving examples in North Wales.

Llanrwst was a large mine marked by an engine house chimney. It is situated at Bwlch yr Haiarn. There are a number of capped and locked shafts. Most prominent is Endean's Engine Shaft next to the old engine house. On a hillock to the south east is the capped Doctor's Shaft. Just across the forest track to the east is the gated Shallow Adit and the capped Air Shaft. The capped Footway and Diagonal

shafts are in the grounds of the Outdoor Centre. To the right of the road to the north, and just opposite the Vale of Conwy Level 2 adit is the gated Llanrwst Deep Adit. Around this an area of extensive surface workings,. They are unfortunately all overgrown, very dangerous and fenced off.

The walk to this mining area also takes in Cyffty. After passing that mine the walk continues along paths and tracks to Llanrwst and the fine chimney.

1. The Llanrwst Mill chimney; 2. The engine house at Llanrwst Mine

Lliwedd Bach Copper Mine

Ordnance Survey 1:25,000 Explorer OL 17
Snowdon/Yr Wyddfa
Grid reference: SH 634 531

Lliwedd was the most productive copper mine in the area. Known sometimes as Bwlch Mwlchan after the farm whose land the mine is situated. The mine was producing copper ore in 1806. After 1821 it was mined by the Lliwedd Mining Company. This was a shaky outfit being comprised of shopkeepers and business men from Ruthin. The majority of the ore was sold at Swansea but some of it was also sold in Liverpool and the Mona works on Anglesey. The mine did not make much profit. A case in example was in 1845 when the best ores of around 12% were realising £8 per ton the ores here were much poorer quality and of the 43 tons sent to Anglesey only £149 was realised. After carriage and royalties there was only £87 left, a meagre profit of £2 per ton! As such closure was inevitable. In 1846 the mine was completely abandoned. In 1853 Lliwedd Mawr mine, this is further up the hillside than Lliwedd Bach, was offered for sale.

1. An entrance into Lliwedd Bach;
2. The ruined flywheel lying broken at Lliwedd Bach; 3. The roller mill with rollers;
4. An overview of the lower mine buildings

1. An entrance into Lliwedd Bach; 2. Upper entrance into the mine; 3. Ruin at Lliwedd Bach above the lower mill area

Eventually it was taken up and was operational again by 1858. The last record of any production was in 1867 when 6 tons were recorded as having been mined. It has been said that this mine 'epitomises the greater extent and spirit of the old men'. It is reliably said that work was carried on here in 1910. No machinery was involved and any ore was picked and washed by hand, placed in sacks and carried down to the port at Porthmadog for shipment to Swansea. The copper was reportedly a bluish colour but with very little sulphur.

There are a number of remains hereabouts and is of great interest. The

actual mine workings are vertical in nature, open-cuts that are almost crevasse like, some .91 to 1.22 metres or 3 to 4 feet wide and of great depth. The lower buildings are the two crushing mills whilst inside there are remains of a pair of fluted crushing rolls. The wheel pit is .91 x 1.82 metres or 30 feet x 6 feet 6 inches that, apparently, was never used. On the ground is an enormous 11 feet or 3.35 metres diameter fly wheel.

The scenic walk to this mine is in high mountainous terrain and weather can change very quickly. Appropriate clothing should be worn and food, water, hot drink should be carried along with sun screen. The entrances are VERY insecure and it inadvisable to go near. Plenty can seen from the footpaths.

Llyn Bodgynydd

Ordnance Survey 1:25,000 Explorer OL 17 Snowdon/Yr Wyddfa
Grid reference: SH 764 595

This mine had several names before Bodgynydd became the accepted one. These were Llynybod and Bodgwynydd whilst another Caegwyn probably related to another mine south of the main mining area of Bodgynydd. The mine was not very productive and was undertaken by several small partnerships between 1849 and 1879. It produced little more than a ton of lead during this period! Possibly, however, the mine was producing lead before the records began in 1849 when it produced a greater quantity as suggested by the extensive but basic dressing floor and two large shafts.

Llyn Bodgynydd is locally known as Llyn Bod or Bod Mawr which distinguishes it from the much smaller lake known as Bod Bach or Cors Bodgynydd reservoir. Water flows from the larger lake into the smaller one.

Both these lakes were dammed to create reservoirs for water to turn the large water wheel at the nearby Pandora mine. Both held much water until the Bod Bach dam was demolished and the sluice control at Bod Mawr altered in 1970. The area beside the smaller lake is now designated as the Cors Bodgynydd Nature Reserve due to the rich variety of plant and animal life there

There remains are not very dramatic but there are a number of small shafts and flooded pools along the Bodgynydd Lode on the southern shore of Llyn Bodgynydd. There is also a fenced off adit to the east of the road, presumably on the same lode.

A circular walk from the National Resources Wales Ty'n Llwyn car park arrives at the mine after a rough walk alongside Llyn Bodgynydd into the Cors Bodgynydd Nature Reserve. The return walk follows the road back to Ty'n Llwyn.

1. The dam on Llyn Bodgynydd; 2. The only remaining ruin of Llyn Bodgynydd; 3. One of several entrance shafts into Llyn Bodgynydd

Llyn Du Manganese Mine

Map: Ordnance Survey 1:25,000 OL 18 Harlech, Porthmadog,& Bala/Y Bala
Grid reference: SH 654 346

The output figures of production for Llyn Du between 1890 and 1897 were possibly linked into the ones for Llyn Eiddewmawr. It is thought that a lorry made the journey here, taking a whole day for the round trip from the station at Talsarnau! Descending from Llyn Du to Llyn Eiddewbach the track is steep and it is probable that pack horses or sledges made this part of the journey. The mine owner, between 1890 and 1892, was the Welsh Manganese Company Ltd., between 1893 and 1895 Alfred Ferguson whilst 1894 to 1897 it was Ellis Pritchard.

The uses of manganese were quite diverse. It was used in the manufacture of bleach; the steel industry; glass making and in the metallurgical and chemical industries. Manganese was mainly used to create stronger steel and beginning in the mid 19th century was used extensively to increase the wear resistance and hardness of steel. This is the largest use today with some 90% of the world production of manganese being used in the iron and steel industries. Around 8 to14lbs (4 to 6Kgs) of manganese are used in every ton of steel depending on the product. Welsh ores were found to be of low quality. As such their main value was as a direct additive to the blast furnace.

The walk to the mine starts from the end of the road beyond Eisingrug. It follows tracks and the old tramway ending at Llyn Du below Moel Ysgyfarnogod 2,044 feet or 623 metres. An ascent to the summit and good viewpoint can be made from the lake.

1. The Llyn Du tramway;
2., 3. & 4. Llyn Du manganese mine ruins

Snowdonia Metal Mines

Llyn y Dywarchen Manganese Mine

Map: Ordnance Survey 1:25,000 Explorer OL 18 Harlech, Porthmadog & Bala/Y Bala Grid reference: SH 657 341

Llyn y Dywarchen workings were less productive than Llyn Eiddew-mawr producing only 2,168 tons and was only worked between 1892 and 1897. It was owned by the Welsh Manganese Company Ltd. The main outcrop is around 230 metres or 250 yards south of Llyn y Dywarchen itself at an altitude of 488 metres or 1,600 feet. It curves round the end of the ridge and then rises towards Llyn Du at a height of 1,800 feet or 549 metres. Generally the bed is 30 centimetres or a foot thick but in places it increases to 50 centimetres to18 inches.

The walk to this mine starts. Following a good track it is found by turning left on a faint path at the start of the tramway to Llyn Du. The remains are reached just before reaching the horseshoe Llyn Dywarchen, a pretty lake.

Llyn y Dywarchen manganese mine

Llyn Eiddew Mawr Manganese Mine

Map: Ordnance Survey 1:25,000 Explorer OL 18 Harlech, Porthmadog & Bala/Y Bala
Grid reference: SH 645 340

The bed of manganese runs south west from the northern point of Llyn Eiddew Mawr for around 183 metres or 600 yards and is about 30 centimetres or a foot thick. It was owned from 1889 to 1891 by the Welsh Manganese Company Ltd then, after a break, by the Cambrian Manganese Company from 1918 to 1922. Workings were mainly opencast and the cutting walls are well preserved. In total the mine produced 3,872 tons from a workforce of 12 in 1889 but only 1 in 1890. Varying small numbers of workers continued mining until 1897 when it closed until 1918 when the highest number of workers was recorded at 14. At closure there were 10 workers.

An easy walk reaches this mine from the end of the road beyond Eisingrug. It follows a good track until just before Llyn Eiddew-bach where a stile over the wall to the right leads to the old workings.

1. The mine site with Llyn Eiddew Mawr beyond; 2. The start of the tramway

Snowdonia Metal Mines

Pandora Lead Mine

Ordnance Survey 1:25,000 Explorer OL 17
Snowdon/Yr Wyddfa
Grid reference: SH 767 602

Llyn Geirionydd is almost a mile long and covers an area of 45 acres with a maximum depth of 15.2 metres or 50 feet. There are few, if any, fish in the lake. This is probably due to the water being poisoned by the water flowing through Pandora Mine from which lead and zinc were extracted. The car park below the mine by the side of Llyn Geirionydd was once the site of the waste tip site for the mine situated above. To the right of the car park is the fenced entrance into the 500 metres or 1,640 feet Pontifex Adit with a small stream issuing out of it. DO NOT enter as the water is cold, waist deep and ends up at a false floor! This adit was used to take ore from the mine to the tramway and thence along this. The tramway was 2.8 kilometres or 1¾ miles long ending at the head of the aerial ropeway taking ore down to Klondyke Mill for processing.

Lead mining started at Pandora in 1870 and over the years was mined by many different companies. Among other names it was known as Willoughby Lead Mine in 1871 then the Welsh Foxdale Lead Mine in 1893 and New Pandora Lead mine in 1908. The last major attempt at mining Pandora was by the Eagle Lead Company in 1920. They continued until 1932. Perhaps the last miner to try their luck here was Evan Thomas prior to 1948. By the time of final closure shortly after, around 3.6 kilometres or 2 miles of levels had been driven since the mine commenced working.

1. The entrance into Pandora; 2. Another entrance; 3. The extensive spoil heap of Pandora; 4. The Pontifex drainage tunnel close to Llyn Geirionydd

There are few remains of the works but across the field an entrance leads into the mine. Specialist equipment is needed to explore this vertical system. It is situated on private land.

Parc Lead Mine including Vale of Conwy Lead Mine

Ordnance Survey 1:25,000 Explorer OL 17 Snowdon/Yr Wyddfa
Grid references: Parc. Lead. No 2 Adit SH 786 598; No 3 Adit SH 787 601; No 5 Adit SH 788 607; Fuchelas Deep Adit SH 786 600; Air Saft and Stopes SH 787 596. Vale of Conwy. Lead. SH 778 596

Parc Mine was the last working mine in the Gwydir Forest and its extensive connections with older mines make it an important resource from the point of view of both history and industrial archaeology. The mine started life as part of Gwydyr Park Consols in 1883 and passed through various hands over the years. While both lead and zinc concentrates were sold, this generally didn't cover working costs of the mines so that many of these enterprises ran at a loss. Eventually the long suffering mine shareholders forced liquidation of the companies and the mine setts were sold on to restart the cycle. After the Second World War prospects improved and more modern equipment and better separation plant increased yields and the mine ran at a profit. Sadly the yields of ore at depth proved to be poor and by the late 50's a combination of low content and poor metal prices meant the enterprise was finished. By that time the principal lode had been driven to connect with the older Llanrwst and Cyffty mines, but neither offered any substantial reserves of ore. During the early 60's the mine was used for experiments with new ore separation techniques and a considerable amount of material was processed.

A very interesting way-marked walk starting from the Hafna Mine car park takes in Hafna Mine, Parc and the Vale of Conwy mines. There is much to see and well worthwhile.

1. Kneebone's Cutting; 2. One of several capped and ventilated shafts into the Vale of Conwy mine; 3. Buddle at Vale of Conwy mine; 4. Ruin of the waterwheel powered mill

Snowdonia Metal Mines   79

Sygun Copper Mine

Ordnance Survey 1:25,000 Explorer OL 17 Snowdon/Yr Wyddfa
Grid reference: SH 605 487

Rocks in the Sygun area are some 450 million years old and were deposited during violent volcanic eruptions. These produced vast quantities of ash which consolidated into Tuff. Cracks in this allowed mineral deposition to take place. Here in particular it was copper, although traces of gold and silver have also been found.

Mining possibly commenced here in Roman times. After they had left the mine remained untouched until the Industrial Revolution when there was a rising demand for copper. In 1836 the annual production was valued at £2,800 and by 1862 some 2 – 3,000 tons of ore had been mined. In 1898 the flotation process increased the output of copper from the ore. This used a mixture of oil and water to extract the copper and was developed by brothers Frank and Stanley Elmore who owned the mine at that time along with their father.

The flotation process entails using the waste material and water. This waste usually has some 1% – 2% of copper ore and is crushed. Water is then added and fed into a ball mill which is a rotating drum with steel balls that crush the ore into a fine powder to form slurry. This is then fed into a rake classifier when it comes out of the mill and then into flotation cells. Any particles that are too large are returned to the mill. Air is injected into the flotation cells where foaming agents are added. This creates froth and copper particles, because of their light weight, become a part of this froth. Heavier particles such as iron sink. The mixture of copper and froth containing around 20% – 40% copper continues for further processing to extract the copper itself.

The most productive period of the mine was during the first half of the 19th century. In 1903 the mine closed as the veins had largely been worked out. In 1983 restoration commenced by the Amies family with a view to open it as a tourist attraction. In 1988 the family was awarded the 'Prince of Wales Award' for the sensitive development of visitor attractions at the mine and was presented by Prince Charles. He was given a gift of an ingot of copper produced from local ore. Nowadays the mine is owned by the Ward family who also refurbished many of the mine attractions.

Interestingly the mine's original buildings were used as a set in the 1958 film 'The Inn of the Sixth Happiness'. This fine show mine, well worth a visit, is signposted beyond Beddgelert when travelling north along the A498.

1. Water wheel at the entrance to the show mine; 2. A mine entrance to the left of the show mine complex; 3. The show mine entrance

Ystrad/Silurian or Garreg Fawr Iron Mine

Ordnance Survey 1:25,000 Explorer OL 17
Snowdon/Yr Wyddfa
Grid reference: SH 544 579

The lower mine was known as the Ystrad or Silurian Mine whilst the upper one was known as Garreg Fawr, also the name of a slate quarry close by.

Prior to 1900 Ystrad Mine was only a small opencast affair. In 1909 it was developed by the Betws Garmon Iron Ore Smelting Company. The mine was connected to the Welsh Narrow Gauge Railway. This is now the Welsh Highland Railway. This company then changed the name to the Phosphoric Iron Ore Company but ran into financial difficulty. It was taken over but only on paper by a new company, the Silurian Iron Ore Company. It used the previous company's assets but did not take on its liabilities and closed in 1919.

Garreg Fawr started out life in the 1840's as a trial for copper but iron ore was mined during the 1840's. The mine was worked in conjunction with the nearby

1. The lower mill area; 2. Ruins at lower level; 3. Looking up to the many levels from the ruins of the mill area

Snowdonia Metal Mines

slate quarry by the Garreg Fawr Slate and Mineral Company in the 1860's. After this period of activity the mine was only worked intermittently by different owners until 1907 when it was taken over by Alfred Hickman Ltd based in Wolverhampton. Major developments took place including the construction of a 4.7 kilometres or 2.9 miles long aerial ropeway. This went up to and over Bwlch y Groes carrying ore to its terminus on the shore of Llyn Padarn in Llanberis where it joined the London and North Western Railway. Closure came in 1913 but during the First World War it briefly re-opened to supplement depleted sources of iron ore.

The walk to this mine follows signed rights of way on paths to reach the lower workings and building remains. It is possible to reach the remainder of the workings that are out of the access land boundary. There are many deep shafts. Do not venture close or enter any of the adits. As parking is difficult the following will help greatly in finding the small parking lay-by to start the walk. From Beddgelert follow the A4085 towards Caernarfon. Drive through Rhyd Ddu and into Betws Garmon. There is a small gravel lay-by beyond Betws Garmon and Salem 50 metres before black and white chevrons indicating a right hand bend and a road narrows sign. From Caernarfon follow the A4085 and drive through Waunfawr and continue towards Rhyd Ddu. Go past Ystrad Isaf, a farm on the left to the speed restriction sign before the second Betws Garmon and the lay-by on the right hand side of the road 150 metres past Yr Hen Ysgol (*The Old School*) and 50 metres beyond a left hand bend.

1. The diagonal line of the workings up the hillside; 2. The drum house below the mill area

1. One of the mine entrances; 2. An inverted V arch to one of the entrances; 3. One of the mine entrances

Snowdonia Metal Mines

Glossary of Terms

ADIT – A horizontal passage in a mine used for drainage, haulage or access. An adit was also driven for exploratory purposes

BARRACKS – The accommodation area used by quarrymen usually through the week but occasionally all year

BLENDE – Zinc ore known as zinc sulphide. Sometimes called BLACKJACK

BUDDLE – This is an elementary, but usually circular ring used for the separation of small sized ore from its associated dirt by means of a stream of water. They are usually shallow but can also be oblong in shape. The ore was placed in the buddle and agitated in a stream of water with the lighter particles being washed away leaving behind the heavier lead or gold

COPPER – A soft reddish mineral and used in alloys such as brass and bronze. Symbol Cu

COPPER PYRITES/CHALCOPYRITE – Also known as yellow sulphide of copper because of the combination of copper and sulphur with an equal proportion of iron. A brassy-yellow unaltered ore that has not been exposed to weathering called Chalcopyrite. Often referred to as 'peacock Ore'. Formula $CuFeS_2$

DRESSING – This is the process of separating the metal ore from the associated gangue, clay or dirt. The end product was ready for smelting

DRESSING FLOOR – Usually refers to sites where semi mechanical or mechanical processes were used

ENGINE SHAFT – A shaft often fitted with pumping equipment

FATHOM – Is a measurement of 6 feet and was the miner's unit of depth

FAULT – A dislocation of rock strata due to tectonic movements

FLAT RODS – Iron rods that were 20 feet long and linked together to transmit power from a waterwheel or rotating engine to the pumping shaft

GALENA – The principal lead ore, sometimes referred to as lead sulphide. Symbol PbS

GANGUE – Other minerals accompanying metallic ores in a vein or deposit

GINGING – The wall like lining at the top of shafts where it passed through loose ground

GUNPOWDER – Explosive black powder used for breaking rock

HORSE WHIM – A windlass that is turned by horse power

IRON PYRITES – Often called 'Fools Gold'. Also simply called pyrite. It is a yellow mineral consisting og of iron sulphide in cubic crystalline form. Symbol FeS

JIGGING – This took place by using a wooden box 1.5 metres or 5 feet or more in length inside which was a sieve and suspended from above by a long pole. The box was then three quarters filled with water and the crude ore placed on the sieve which was then agitated by jerking up and down in the water by means of the pole. Repeated actions separated the lighter 'gangue' minerals from the heavier ore with the lighter minerals being skimmed off

KIBBLE – An iron or iron hooped bucket used for bringing the crude ore up a shaft

LEAD – A toxic bluish white metallic element used on it own or in alloys, cable sheaths, paints and as radiation shields. Symbol Pb

LAUNDER PILLARS – These pillars were built from stone and were used to support a trough, often wooded, that allowed water to be conveyed to a waterwheel or for the buddles

LEAT – An artificial watercourse or aqueduct dug into the ground, especially one supplying water to a mill

LEVEL – A horizontal passage in a mine and used for drainage haulage, ventilation or access

LODE – A vein or rake of mineral in a fissure

MALACHITE – A very rich ore of copper carbonate Formula $Cu_2CO_3(OH_2)$

MILL – The building where the mineral is reduced by using machinery

OCHRE – A name given to oxides of iron and manganese and has a range of colours, black, brown, orange and red

ORE – The mined material form which ore can be extracted

RISE – A shaft driven upwards from a level

SETT – An area of land leased for mining

SHAFT – A vertical or nearly so opening into a mine. They were used for a number of purposes that included access, winding or hauling out the ore and ventilation

SILVER – Associated with lead. Generally around 2 ounces per ton were extracted

SMELTER – The furnace for processing copper and lead ores. It is possible that the smelting process was carried on 6000 years ago in Iran

STAMPS – These have long been associated with ore crushing even as early as 1553

STOPE – Part of a lode where ore is removed from the lode or vein

TAKE NOTE – A lease that allows the mining of a sett

TRIAL – Exploratory tunnel

WATERWHEELS – These were used as a source of power for a variety of purposes not least for powering crushing machinery. A small working wheel can be seen at Sygun Copper Mine and a much larger one in Cwm Ciprwth

WINZE – A shaft that has been sunk below a level

ZINC – Sometimes called Sphalerite A brittle bluish white mineral that is used in alloys such as brass and as a coating to form a protective coating on metals such as iron and as battery electrodes. Symbol Zn

Selected Bibliography

MINES OF THE GWYDYR FOREST Part 1 – John Bennett and Robert Vernon. This volume covers the Llanrwst, Alltwen and Gorlan areas.
ISBN 0 9514798 0 6

MINES OF THE GWYDYR FOREST Part 2 – John Bennett and Robert Vernon. This volume covers the Hafna and Vale of Conwy areas.
ISBN 0 9514798 1 4

MINES OF THE GWYDYR FOREST Part 3 – John Bennett and Robert Vernon. This volume covers the Parc, Tyn Twll and Gwydyr Consols areas.
ISBN 0 9514798 2 2

MINES OF THE GWYDYR FOREST Part 4 – John Bennett and Robert Vernon. This volume covers the Aberllyn, Penrallt, Griffin and Wheal Gorge areas.
ISBN 0 9514798 3 0

MINES OF THE GWYDYR FOREST Part 5 – John Bennett and Robert Vernon. This volume covers the Cyffty, Coed Mawr Pool, Ffrith and Glyn areas.
ISBN 0 9514798 4 9

MINES OF THE GWYDYR FOREST Part 6 – John Bennett and Robert Vernon. This volume covers the Pandora, Klondyke and Caerhegla areas.
ISBN 0 9514798 5 7

MINES OF THE GWYDYR FOREST Part 7 – John Bennett and Robert Vernon. This volume covers the Cae Coch, Coed Gwydyr and Trecastell areas.
ISBN 0 9514798 6 5

THE OLD COPPER MINES OF SNOWDONIA – David Bick.
ISBN 0 906885 03 5

1. Looking up to Hafod y Llan copper mine;
2. The diagonal line of the Ystrad/Silurian or Garreg Fawr Fron mine; 3. Ruin at Hafod y Porth

COMPACT CYMRU
– MORE TITLES:

www.carreg-gwalch.cymru

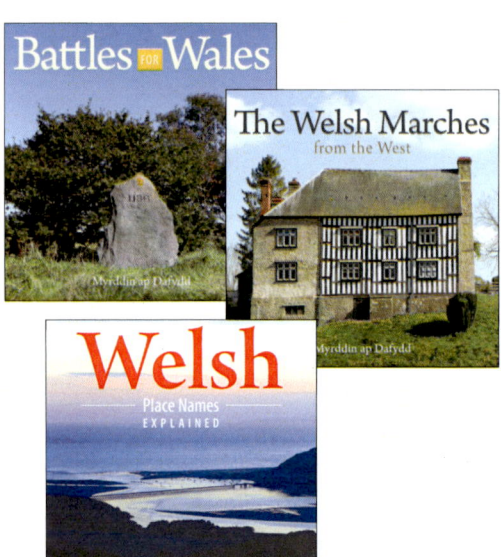

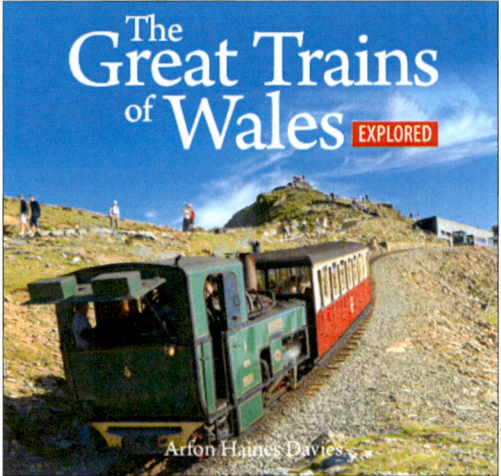

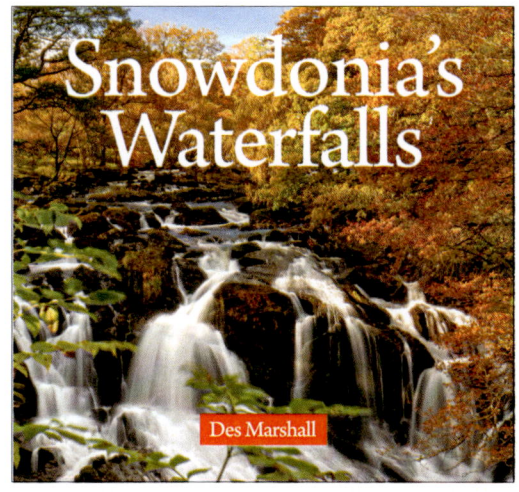

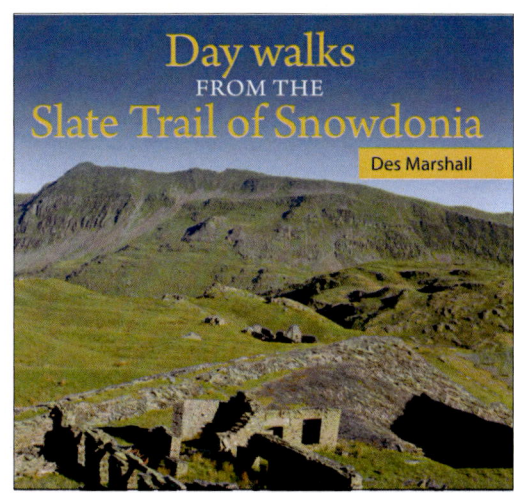

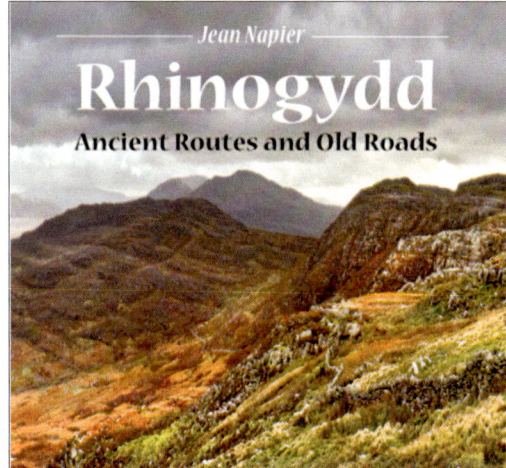

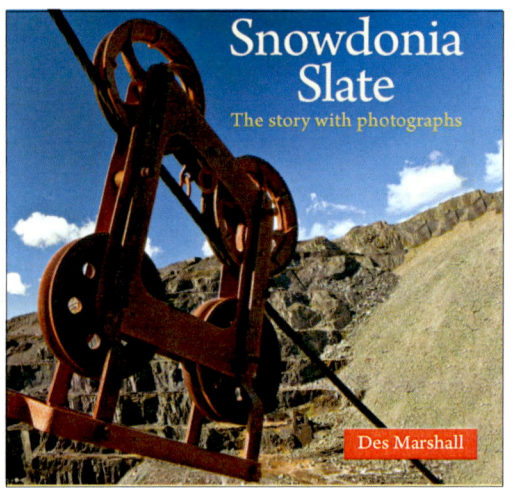
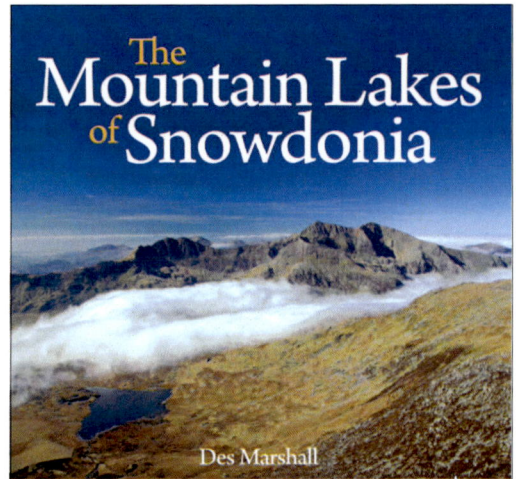
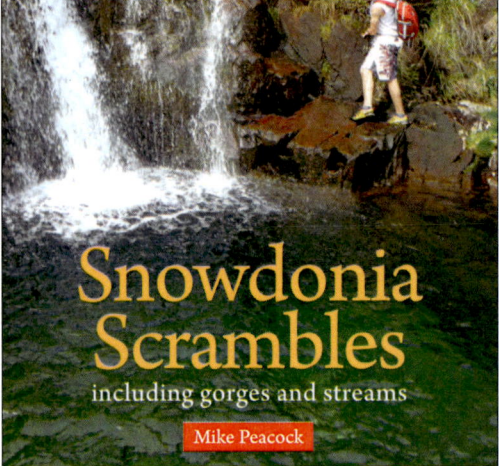
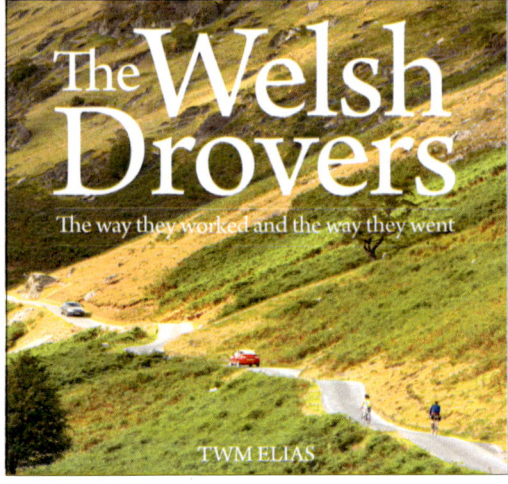